HOW TO PASS

CW00494805

FLASH

HIGHER BIOLOGY

Billy Dickson
Graham Moffat

HODDER GIBSON
AN HACHETTE UK COMPANY

Orders: please contact Bookpoint Ltd, 130 Milton Park, Abingdon, Oxon OX14 4SB. Telephone: (44) 01235 827720. Fax: (44) 01235 400454. Lines are open 9.00–5.00, Monday to Saturday, with a 24-hour message answering service. Visit our website at www.hoddereducation.co.uk. Hodder Gibson can be contacted direct on: Tel: 0141 848 1609; Fax: 0141 889 6315; email: hoddergibson@hodder.co.uk

First published in 2011 by
Hodder Gibson, an imprint of Hodder Education,
An Hachette UK Company
2a Christie Street
Paisley PA1 1NB

Impression number 5 4 3 2 1
Year 2013 2012 2011

Cover photo © Photodisc/Getty Images; © iStockphoto.com
Illustrations by GreenGate Publishing Services
Typeset in Gill Sans Alternative by GreenGate Publishing Services, Tonbridge, Kent
Printed in Great Britain by CPI Antony Rowe

A catalogue record for this title is available from the British Library

ISBN: 978 1444 120493

Contents

Contents

Growth and regulation

Control of growth and development

Physiological homeostasis

Populations

Cell variety (1)

The diagrams show the detailed structure of three different animal cells.

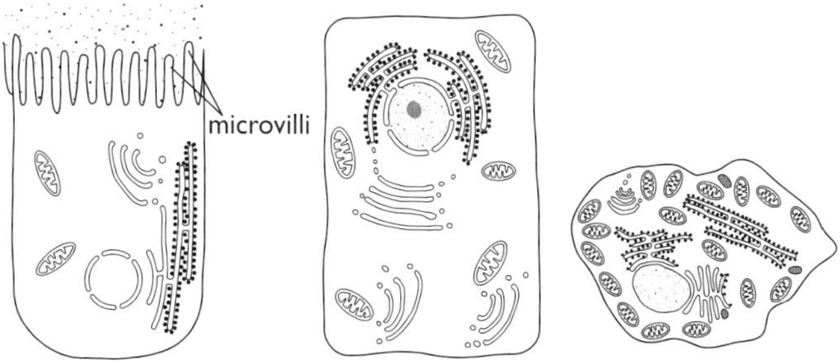

1 2 3

Q1 **Diagram 1** shows a cell from the lining of the small intestine of a mammal. Suggest a reason for the presence of microvilli in this type of cell.

Q2 **Diagram 2** shows a cell from the pancreas. Name the structures shown in the diagram which are involved in the secretion of insulin.

Q3 **Diagram 3** shows a liver cell. Identify a feature of the cell which indicates that it has a high energy demand.

Q4 Give a feature of a red blood cell that increases the efficiency with which it absorbs and releases oxygen.

Q5 Name the cell structures found in phagocytes which contain the digestive enzymes involved in the destruction of bacteria.

ANSWERS

1 The microvilli provide a large surface area for the absorption/diffusion of food molecules.

2 The presence of Golgi apparatus *and* secretory vesicles.

3 The presence of a large number of mitochondria.

4 It has a large surface area *or* it contains haemoglobin which carries oxygen.

5 lysosomes

***Exam* tip:** Make sure you can identify all the structures of an animal cell and state their functions. Remember to *link* these structures to the specialised function of the cell.

Cell variety (2)

The diagrams show the detailed structure of two different plant cells.

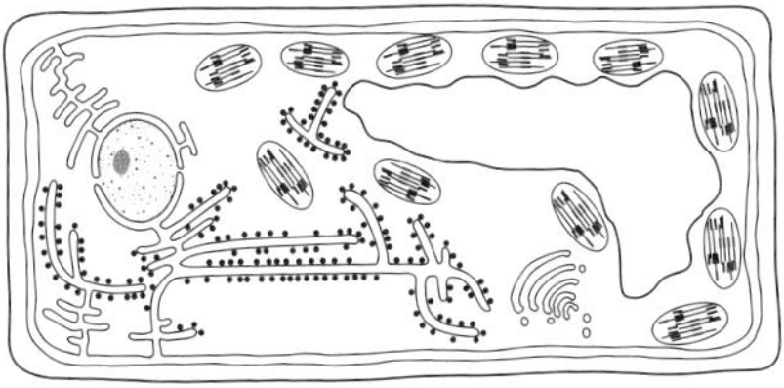

1

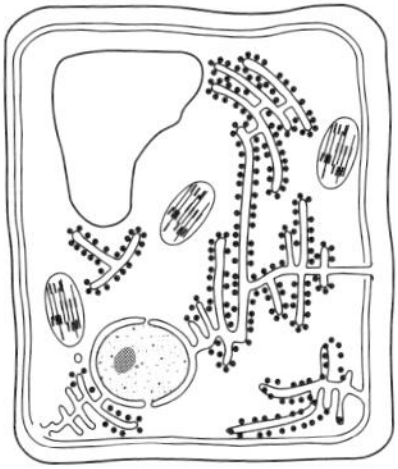

2

Q1 **Diagram 1** shows a cell from the mesophyll of a green leaf. Identify a feature of the cell which indicates that it is involved in photosynthesis.

Q2 **Diagram 2** shows an undifferentiated plant cell. Identify a feature of the cell which indicates that it is involved in the synthesis of protein.

Q3 Describe a structural adaptation of a root hair cell which increases its efficiency in the absorption of water and minerals.

Q4 Give a feature of xylem vessels which contribute to their role in supporting plant stems.

Q5 Name the specialised cells whose changes in turgor cause the opening and closing of stomata.

ANSWERS

1 The presence of a large number of chloroplasts.

2 The presence of a large number of ribosomes.

3 Cell wall projection/hair shape increases the surface area of the cell.

4 Their walls are thickened with lignin.

5 guard cells

***Exam* tip**: Make sure you can identify all the structures of a plant cell and state their functions. Remember to *link* these structures to the specialised function of the cell.

Unicellular organisms (1)

Q1 Give the meaning of the term unicellular organism.

The diagram shows a yeast cell.

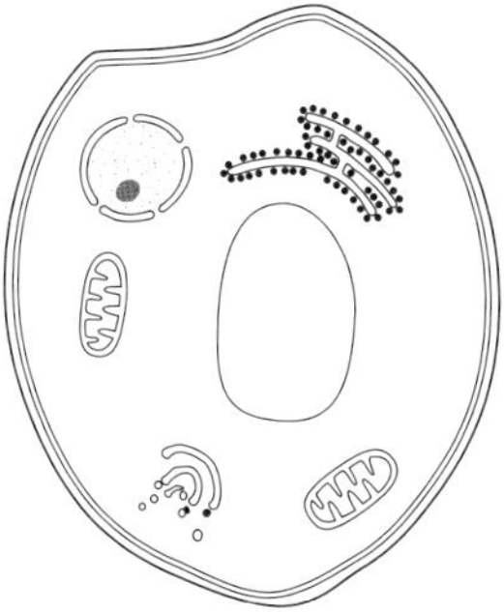

Yeast cell

Q2 What evidence from the diagram supports the statement that yeast cells cannot produce their own food by photosynthesis?

Q3 What feature of the yeast cell suggests that it is *not* an animal cell?

Q4 What evidence is there that yeast cells can secrete enzymes?

Q5 Yeast cells can respire anaerobically. What evidence is there that they can also carry out aerobic respiration?

ANSWERS

1 An organism that has only one cell and contains all the organelles it needs to survive.

2 Yeast cells do not contain chloroplasts.

3 The presence of a cell wall or large vacuole.

4 The presence of Golgi apparatus and secretory vesicles.

5 The presence of mitochondria.

***Exam* tip**: You should be able to apply your knowledge of cell structures and organelles from *all parts of the Course* to explain their roles in a unicellular organism.

Unicellular organisms (2)

The diagrams show two unicellular organisms.

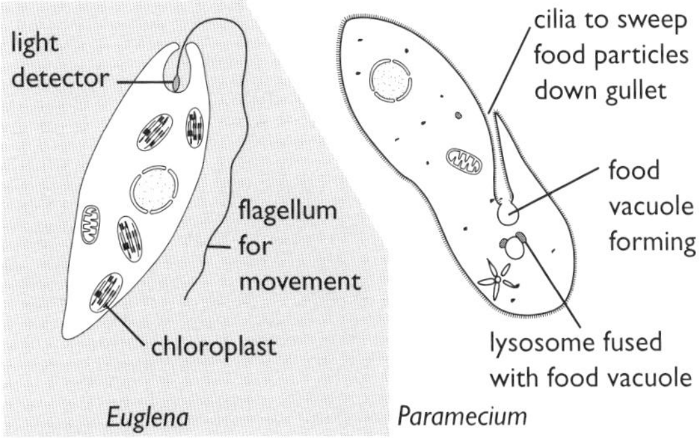

Q1 Which feature of *Euglena* shows that it can carry out photosynthesis?

Q2 Explain how the structure of *Euglena* shown in the diagram allows it to photosynthesise efficiently.

Q3 Use evidence from the diagrams to explain how *Paramecium* obtains its food.

Q4 Use evidence from the diagrams to explain how *Paramecium* digests its food.

Q5 Describe the process of gas exchange in **both** *Euglena* and *Paramecium*.

ANSWERS

1 The presence of chloroplasts.

2 It can detect light using its light detector *and* move towards it using its flagellum.

3 It moves food towards and into its gullet using cilia then engulfs the food into vacuoles.

4 The lysosomes fuse with a food vacuole and provide the enzymes to digest the food.

5 In both organisms gases are exchanged by diffusion through their cell membranes.

***Exam* tip:** This is another opportunity to apply your knowledge of cell structures and organelles from *all parts of the Course* to explain their roles in a unicellular organism.

Cell structure related to function

The diagram represents a cell from a human salivary gland.

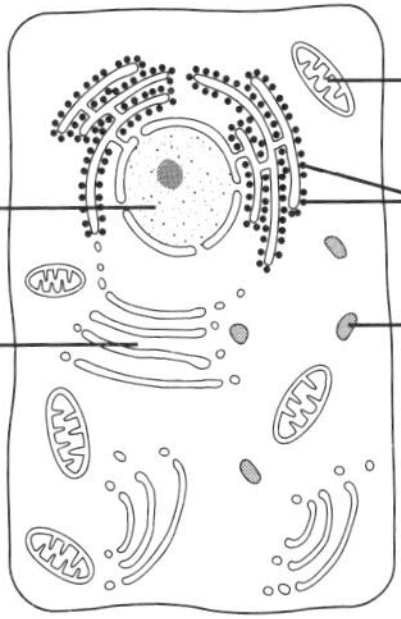

Q1 Name the structure which is the site of mRNA synthesis.

Q2 State the functions of the ribosomes, rough endoplasmic reticulum and Golgi apparatus.

Q3 Give the function of mitochondria.

Q4 Name the structure which has ribosomes attached to its surface.

Q5 Describe the function of the lysosomes.

ANSWERS

1 nucleus

2 Ribosomes are the sites of protein synthesis; rough endoplasmic reticulum is an internal transport network for proteins; Golgi apparatus packages and processes protein for secretion.

3 Mitochondria are the sites of aerobic respiration.

4 rough endoplasmic reticulum (ER)

5 Lysosomes contain digestive enzymes that break down unwanted cell materials.

***Exam* tip:** Ribosomes and rough ER are often asked about in relation to cells that produce enzymes or hormones; mitochondria in relation to muscle or liver cells; lysosomes in relation to phagocytes.

Diffusion

Q1 Describe the process of diffusion.

Q2 Explain the importance of diffusion to living cells.

Q3 Explain why starch molecules cannot diffuse through cell membranes.

Q4 Explain why the uptake of ions by diffusion into a cell is *not* affected by a change from aerobic to anaerobic conditions.

Q5 Describe what is meant by the term concentration gradient.

ANSWERS

1 The movement of particles from a region of high concentration to a region of low concentration until they are evenly spread.

2 Diffusion allows glucose and oxygen to enter a cell for respiration, which releases energy for other cell processes.

or Diffusion allows amino acids to enter a cell for enzyme synthesis, which allows cell metabolism to continue.

or Diffusion allows the removal of wastes such as carbon dioxide which could be harmful to the cell if they were allowed to build up.

or Diffusion allows carbon dioxide to enter a plant cell for the process of photosynthesis.

3 Starch molecules are too large to pass through the cell membrane.

4 Diffusion is a passive process that does not require energy from aerobic respiration.

5 Concentration gradient describes the difference in concentration of a particular substance between two locations.

***Exam* tip:** Remember the importance of diffusion to cells or unicellular organisms as GORE – Glucose and Oxygen for Respiration for Energy.

Remember the importance of *time* for diffusion to occur.

Osmosis

Q1 Describe the process of osmosis.

Q2 Describe what is meant by the terms hypertonic, hypotonic and isotonic solutions.

Q3 Give the terms used for solutions described by the following. It:

- **a** has a higher water concentration than a solution on the other side of a membrane
- **b** has a lower water concentration than a solution on the other side of a membrane
- **c** could make a plant cell plasmolysed or flaccid
- **d** could make an animal cell shrink
- **e** could make a plant cell turgid
- **f** could make an animal cell swell and burst.

Q4 Describe the appearance of a plant cell immersed in a hypotonic solution.

Q5 Describe the appearance of a plant cell immersed in a hypertonic solution.

ANSWERS

7 Osmosis

1 Osmosis is the movement of water molecules from a region of high water concentration to a region of lower water concentration through a selectively permeable membrane.

2 Hypertonic means a solution or tissue with a lower water concentration compared with another.

Hypotonic means a solution or tissue with a higher water concentration compared with another.

Isotonic means a solution or tissue with the same water concentration compared with another.

3 **a** hypotonic

b hypertonic

c hypertonic

d hypertonic

e hypotonic

f hypotonic.

4 Cell is turgid with a large swollen central vacuole forcing the cell walls to bulge out.

5 Cell is plasmolysed with a small shrunken central vacuole which has pulled the cell membrane away from the cell wall in places.

***Exam* tip:** In describing osmotic effects, always state the difference in water concentration and state which way water moves. Finish by describing the effect that the water movement has had on the cell or tissue.

Active transport

Q1 Describe the process of active transport.

Q2 Explain the effect of temperature on the rate of active uptake of ions by plant root cells.

Q3 Explain why the uptake of nitrate ions by barley roots would increase as the concentration of oxygen increases.

Q4 When oxygen was bubbled through a tank of unicellular algae, the potassium ion concentration within the organisms increased.

Explain this effect.

Q5 Explain the expected effect of the presence of lead salts on the uptake of other ions by plant roots.

ANSWERS

1 Active transport is the movement of particles from a region of low concentration of the particles to a region of higher concentration. This requires energy.

2 As temperature increases, the rate of activity of enzymes involved in respiration increases, resulting in more energy available for active transport. At higher temperatures enzymes become denatured, stopping respiration and giving no energy for active transport.

3 Increased oxygen can result in increased aerobic respiration and the production of more energy, which could increase the rate of active transport.

4 The oxygen allowed more respiration, which in turn produced more energy. The increased energy resulted in more active transport, which is the main method of uptake of ions such as potassium.

5 Lead salts act as inhibitors of enzymes and so slow down the rate of reactions such as respiration. Less respiration means less energy for active transport, causing a decrease in the rate of uptake of ions by the roots.

***Exam* tip:** Active transport questions are often worth 2 marks.

Remember to include the **CORE** facts in every answer.

Concentration gradient; **O**xygen; **R**espiration; **E**nergy/ATP.

Plasma membrane

Q1 Describe the structure of the plasma membrane.

Q2 State the function of the phospholipid and protein molecules in a plasma membrane.

Q3 Give the overall function of plasma membranes.

Q4 Explain the term fluid-mosaic model of the plasma membrane.

Q5 State the properties of plasma membranes that result from the phospholipid and protein arrangement.

ANSWERS

1 The membrane is composed of a double layer of phospholipid molecules with a patchy arrangement of protein molecules throughout it.

2 The phospholipids give the membrane flexibility, while the proteins have special functions such as the creation of pores to give the membrane selective permeability or as carriers that allow active transport.

3 Cell membranes control the entry and exit of substances into and out of the cell.

4 The fluid part of the fluid-mosaic model refers to the flexibility of the phospholipid bilayer, while the mosaic refers to the patchy arrangement of protein throughout it.

5 The phospholipids give the membrane flexibility, while the channel-forming proteins make the membrane porous and lead to its selectively permeable nature.

***Exam* tip:** Remember the **3 Fs** and the **3 Ps**.

3Fs = (**F**) phospholipids, **F**luid and **F**lexible

3Ps = **P**roteins, **P**atchy and **P**orous

Cell wall

Q1 Name the substance of which the cell wall is composed.

Q2 Describe the structure of the cell wall.

Q3 Describe the permeability of the cell wall.

Q4 Describe the general functions of the cell wall.

Q5 Describe the function of the cell wall in turgidity.

ANSWERS

1 cellulose

2 Cellulose is arranged as fibres.

3 Wall is fully permeable.

4 Walls give strength and support to plant cells.

5 Walls prevent bursting of cells due to entry of water.

Exam **tip**: For **structure** always use the phrase *cellulose fibres*.

For **permeability** use *freely* or *fully permeable*.

Leaves and light

Q1 Give *three* possible fates of light striking a leaf.

Q2 Give the term which applies to light that is taken into the leaf cells.

Q3 Give the term which is applied to light that is bounced back from the leaf surface.

Q4 Give the term which applies to light that passes through the leaf.

Q5 State *one* use of light that is absorbed by leaf cells.

ANSWERS

1 Absorbed, reflected and transmitted.

2 absorption

3 reflection

4 transmission

5 Used in photosynthesis/photolysis/ATP production.

***Exam* tip:** For the fate of light striking a green leaf remember **ART**.

Absorption, **R**eflection and **T**ransmission.

Spectra

Q1 Describe what is meant by a spectrum of light.

Q2 State what is meant by the absorption spectrum of chlorophyll a.

Q3 State what is meant by an action spectrum of a plant.

Q4 Explain why there are differences between the absorption spectrum of chlorophyll a and the action spectrum of a whole plant.

Q5 State the regions of the absorption spectrum for chlorophyll a in which the greatest percentage of available light is absorbed.

ANSWERS

1 A spectrum is a range of wavelengths of light.

2 It refers to the range of wavelengths of light absorbed by chlorophyll a.

3 An action spectrum refers to the rate of photosynthesis carried out at each wavelength of light by a plant.

4 At certain wavelengths of light, the rate of photosynthesis of a plant may be high even though the absorption of that wavelength by chlorophyll a is low. This shows the presence of accessory pigments in plants.

5 The red and blue regions.

***Exam* tip:** Questions on this section often include graphs which require you to select information. Try to use evidence and values with the appropriate units to support your answer.

Pigments

Q1 Name the main photosynthetic pigment and the accessory pigments.

Q2 State what happens to the light energy absorbed by the accessory pigments.

Q3 Name the colours of light that are mainly absorbed by chlorophyll a.

Q4 State the benefit to plants of having accessory pigments.

Q5 State the exact location of the pigments that absorb light within leaf cells.

ANSWERS

1 The main photosynthetic pigment is chlorophyll a and the accessory pigments are chlorophyll b, xanthophyll and carotene.

2 The light energy absorbed by the accessory pigments is passed on to chlorophyll a.

3 red and blue

4 Accessory pigments increase the range of wavelengths/colours of light absorbed and so more energy is available for photosynthesis.

5 The grana of the chloroplasts.

***Exam* tip:** The role of the accessory pigments is often poorly answered. Make sure that you learn the response given in Answer 4.

Chromatography

Q1 Name a technique that is used to separate photosynthetic pigments.

Q2 Give examples of the solvents used to extract photosynthetic pigments from leaf cells and to separate them in chromatography.

Q3 State the order of separation of the pigments on chromatography paper with the least soluble in the solvent first.

Q4 Explain why the pigment spot loaded onto chromatography paper should be above the solvent level and be densely applied.

Q5 Occasionally results are obtained in which the pigments are all carried to the top of the chromatography paper and so fail to separate.

Describe an error in experimental procedure that could account for this.

ANSWERS

1 chromatography

2 Extraction with acetone or propanone and separation with an acetone and petroleum ether mixture.

3 Chlorophyll b, chlorophyll a, xanthophyll then carotene.

4 The pigment should be above the solvent level so that the mixture does not dissolve directly into the solvent but is carried up the paper as the solvent rises. The denser the spot, the more visible the individual pigment marks on the paper.

5 Leaving the separation to run for too long.

***Exam* tip:** Remember the order of the pigments as they separate from the bottom of the paper as **bax c**.

Light-dependent stage (1)

Q1 State the exact location of the light-dependent stage of photosynthesis within a leaf.

Q2 Name the stage in photosynthesis in which light energy is converted into chemical energy.

Q3 Describe how light energy is converted into chemical energy in photosynthesis.

Q4 Apart from ATP, name another product of the light-dependent stage which is required in the carbon fixation stage.

Q5 Name the process in which water is split during the light-dependent stage of photosynthesis.

ANSWERS

15 Light-dependent stage (1)

1 The grana of the chloroplasts.

2 light-dependent stage

3 Light energy trapped by chlorophyll is then used to add Pi to ADP to produce ATP.

4 NADPH

5 photolysis

***Exam* tip:** NAD**P** is the hydrogen acceptor in **P**hotosynthesis. Candidates sometimes get it confused with NAD, a hydrogen carrier in respiration.

Light-dependent stage (2)

Q1 Describe how hydrogen is produced in the light-dependent stage of photosynthesis.

Q2 Name the carrier that transfers hydrogen from the light-dependent stage to the carbon fixation stage.

Q3 Name the by-product produced by the light-dependent stage.

Q4 Explain why a low light intensity may limit the rate of photosynthesis.

Q5 The rate of photosynthesis can be calculated by measuring the change in dry mass with time.

State *one* other method that can be used to calculate the rate of photosynthesis.

ANSWERS

1 It is produced from the splitting of water molecules.

2 NADP

3 oxygen

4 Low light intensity may limit the quantity of ATP and NADPH produced and so limit the carbon fixation stage.

5 Measure the volume of oxygen produced per minute by a photosynthesising plant.

***Exam* tip:** Candidates often struggle with Answer 5.

Any question on rates *must* include a measurement of time.

Chloroplast structure and function

Q1 Describe the structure of a chloroplast.

Q2 Name the part of a chloroplast that contains the pigments responsible for the absorption of light energy.

Q3 Name the part of the chloroplast which contains enzymes that control the reactions of the carbon fixation stage.

Q4 State the exact location of the light-dependent stage within a chloroplast.

Q5 State the exact location of the carbon fixation stage within a chloroplast.

ANSWERS

1 A chloroplast is surrounded by a double membrane. Inside are a fluid-filled region called the stroma and stacks of membranes called grana.

2 granum/grana

3 stroma

4 granum/grana

5 stroma

***Exam* tip:** A simple but well-labelled *diagram* of a chloroplast can earn you useful marks in extended response questions.

Carbon fixation stage (1)

Q1 State the exact location of the carbon fixation stage in a leaf cell.

Q2 Name the *two* products of the light-dependent stage that are required for the conversion of GP into glucose.

Q3 Name the carbon dioxide acceptor in the carbon fixation stage.

Q4 Describe the role of ATP in photosynthesis.

Q5 Describe the role of hydrogen and ATP in the carbon fixation stage.

ANSWERS

1 The stroma of the chloroplasts.

2 ATP and NADPH

3 RuBP

4 Transfers chemical energy from the light-dependent stage to allow the reduction of GP to glucose.

5 Hydrogen is required to reduce carbon dioxide to carbohydrate; ATP provides energy for the reduction of carbon dioxide to carbohydrate.

***Exam* tip:** Many candidates lose marks because they fail to use the correct phrases or terminology needed to demonstrate full understanding. Answers 4 and 5 are examples worth learning.

Carbon fixation stage (2)

Q1 Describe the relationship between GP and RuBP.

Q2 State the numbers of carbon atoms in RUBP, GP and glucose.

Q3 Explain the effect on RUBP concentration if conditions were changed from light with carbon dioxide present to light with carbon dioxide absent.

Q4 Explain why a low temperature may limit the rate of photosynthesis.

Q5 Other than carbohydrates, name *two* major biological molecules that are derived from the photosynthetic process.

ANSWERS

1 RuBP accepts carbon dioxide and is converted to GP.

2 5C in RuBP; 3C in GP and 6C in glucose.

3 The removal of carbon dioxide decreases the amount of GP that can be made, so that RuBP would increase.

4 The carbon fixation stage is controlled by enzymes that are temperature dependent, so reducing temperature could limit the rate of photosynthesis.

5 Protein, lipid, nucleic acids (any two).

***Exam* tip**: You must know the numbers of carbon atoms per molecule of carbon dioxide, RuBP, GP and glucose. You should also try to understand the relationship between the RuBP and GP and so be able to interpret graphs showing this.

ATP

Q1 Describe the importance of ATP in cells.

Q2 The quantity of ATP in the human body remains relatively constant, yet ATP is continually being broken down.

Explain this observation.

Q3 Give an example of a process in muscle cells that requires energy released from ATP.

Q4 State the form of energy transferred by ATP.

Q5 Name the stage of respiration that releases the most energy for use by cells.

ANSWERS

1 ATP transfers chemical energy from energy-releasing reactions to cellular processes that require energy.

2 ATP is regenerated at the same rate as it is broken down.

3 Muscle fibre contraction *or* protein synthesis *or* glycolysis *or* DNA replication.

4 chemical energy

5 The cytochrome system.

***Exam* tip:** Pay particular attention to Answers 1 and 2. These responses are seldom given by candidates.

Glycolysis (1)

Q1 State the exact location of glycolysis in a cell.

Q2 Apart from glucose and enzymes, name a substance that is essential for glycolysis to occur.

Q3 Other than carbohydrate, name an alternative respiratory substrate.

Q4 State the net gain of ATP molecules from the breakdown of a glucose molecule in glycolysis.

Q5 Explain why the gain of ATP in glycolysis is described as a *net* gain.

ANSWERS

21 Glycolysis (1)

1 The cytoplasm.

2 ATP

3 Protein *or* amino acids *or* fat *or* fatty acid *or* glycerol.

4 There is a net gain of two ATP molecules.

5 Net gain is an overall gain with some loss but more gain *or* two ATP molecules are used in glycolysis but four ATP molecules are produced.

***Exam* tip:** In this section, the terms *alternative respiratory substrate* and *net gain of ATP* need to be clearly understood.

Glycolysis (2)

Q1 Name the usual respiratory substrate used in glycolysis.

Q2 Name the product of glycolysis that is required by Krebs cycle.

Q3 Name the product of glycolysis that is passed to the cytochrome system.

Q4 Identify which of the following statements about glycolysis is true.

A Carbon dioxide is released.
B Oxygen is required.
C The final respiratory product is lactic acid.
D It is anaerobic.

Q5 State the number of carbon atoms in each molecule of glucose and pyruvic acid.

ANSWERS

1 glucose

2 pyruvic acid

3 NADH

4 D

5 Glucose contains six carbon atoms per molecule and pyruvic acid contains three carbon atoms per molecule.

***Exam* tip:** In extended response questions, candidates often forget to state that oxygen is *not* required for glycolysis and that glycolysis is an *enzyme-controlled* reaction.

Krebs cycle (1)

Q1 State the exact location of the Krebs cycle in a cell.

Q2 Give the conditions required for the Krebs cycle to occur.

Q3 Name the molecule that carries the acetyl group to the Krebs cycle.

Q4 Name the six-carbon molecule produced in the Krebs cycle when an acetyl group combines with a four-carbon molecule.

Q5 Name the substances that control the sequence of reactions in the Krebs cycle.

ANSWERS

1 The matrix of the mitochondria.

2 aerobic conditions

3 co-enzyme A

4 citric acid

5 enzymes

***Exam* tip**: The role of co-enzyme A (CoA) at this stage is often poorly understood – it simply carries the acetyl group into the Krebs cycle.

Krebs cycle (2)

Q1 Name the *two* final metabolic products released during the Krebs cycle.

Q2 Name the carrier substance that transports hydrogen from the Krebs cycle to the cytochrome systems.

Q3 State the number of carbon atoms present in the acetyl group, a citric acid molecule and in a carbon dioxide molecule.

Q4 Suggest why the reactions of the Krebs cycle are temperature dependent.

Q5 Identify which of the following statements is true of the Krebs cycle.

A It requires oxygen.
B Water is a final metabolic product.
C Carbon dioxide is required.
D It is anaerobic.

ANSWERS

1 ATP and carbon dioxide.

2 NAD

3 The acetyl group contains two carbon atoms; a citric acid molecule contains six carbon atoms, and a carbon dioxide molecule contains one carbon atom.

4 They are controlled by enzymes.

5 A

***Exam* tip:** The carrier NAD gains hydrogen and is said to be reduced. Remember **Oil Rig** in Biology – **O**xidation **i**s **l**oss of hydrogen, **R**eduction **i**s **g**ain of hydrogen.

The cytochrome system

Q1 State the exact location of the cytochrome system in a cell.

Q2 Explain why the cytochrome system cannot function in anaerobic conditions.

Q3 Describe the role of oxygen in the cytochrome system.

Q4 State the number of ATPs produced by the cytochrome system.

Q5 Name the *two* final metabolic products of the cytochrome system.

ANSWERS

1 The cristae of the mitochondria.

2 Oxygen is required as the final hydrogen acceptor.

3 Combines with hydrogen to form water *or* is the final hydrogen acceptor.

4 Thirty-six ATPs per molecule of glucose metabolised.

5 ATP and water

Exam **tip:** More ATP molecules are produced during this stage of respiration than at any other.

Mitochondria

Q1 State the function of the mitochondria.

Q2 Describe the structure of a mitochondrion.

Q3 Name the stage(s) of respiration that takes place in the central matrix of the mitochondria.

Q4 Name the hydrogen carrier system present on the cristae of the mitochondria.

Q5 Explain the advantage to muscle cells of having more cristae in their mitochondria than other types of cell do.

ANSWERS

1 They are the sites of aerobic respiration in cells.

2 A mitochondrion has an outer membrane and an inner membrane folded to form cristae. The fluid-filled centre is called the matrix.

3 Krebs cycle

4 cytochrome system

5 This gives a large surface area for increased production of ATP in energetic cells such as muscle cells.

***Exam* tip**: Answer 5 is an example of a less familiar aspect of this topic – worth marks though!

Anaerobic respiration

Q1 State what is meant by the term anaerobic respiration.

Q2 State the location of anaerobic respiration in a cell.

Q3 Name the end products of anaerobic respiration in plant and animal cells.

Q4 Other than their end products, state *one* difference between anaerobic respiration in animal cells compared with plant cells.

Q5 Compare anaerobic and aerobic respiration in terms of the ATP produced.

ANSWERS

27 Anaerobic respiration

1 Respiration in the absence of oxygen.

2 cytoplasm

3 Plants: carbon dioxide and ethanol.
Animals: lactic acid.

4 In animal cells anaerobic respiration is reversible.

5 Aerobic respiration produces more molecules of ATP (38) for every molecule of glucose respired than anaerobic respiration (2).

***Exam* tip:** Candidates should be able to describe the differences and similarities between anaerobic respiration in plants and animals and compare the ATP production from aerobic and anaerobic respiration.

Variety of proteins

Q1 Name the building blocks of protein molecules.

Q2 Name the type of bond that links the building blocks of a protein molecule.

Q3 Explain the importance of the sequence of amino acids in a protein.

Q4 Explain why enzyme molecules can be described as specific.

Q5 Name the class of protein to which the following belong:

a cellulase
b collagen.

ANSWERS

1 amino acids

2 peptide bond

3 Gives the protein molecule its three-dimensional shape and so determines its function.

4 Their three-dimensional shape allows them to act on one type of substrate only.

5 a globular

b fibrous

***Exam* tip:** You need to know that proteins can be classified as either fibrous or globular. Collagen is the only example of a fibrous protein that you can be asked to recall in the exam. The globular proteins include enzymes, hormones and antibodies.

DNA structure

Q1 Nucleic acid molecules are made up of nucleotides.
Name the components of a nucleotide.

Q2 Name the four nucleotide bases in DNA.

Q3 Name the type of bond that joins complementary bases in DNA.

Q4 Give the term used to describe the coiled structure of a DNA molecule.

Q5 Give the term used to describe a sequence of DNA bases that code for one type of protein.

ANSWERS

1 Phosphate, deoxyribose sugar and (nitrogenous) base.

2 Adenine, thymine, guanine and cytosine.

3 hydrogen

4 double helix

5 A gene.

***Exam* tip:** In your exam, you must be able to *name* the bases – not just give their letters. You must not shorten deoxyribose sugar to sugar alone.

DNA replication

Q1 Name the site of DNA replication.

Q2 Name *four* substances needed for DNA replication to occur.

Q3 Describe the first event in DNA replication.

Q4 Place the following events occurring during DNA replication in the correct order:

A Sugar-phosphate bonds form.
B Hydrogen bonds break.
C Complementary base pairing occurs.

Q5 Explain the importance of DNA replication to cells.

ANSWERS

1 nucleus

2 Original DNA template, ATP, enzymes, free DNA nucleotides.

3 Original DNA molecule untwists and unzips.

4 B, C, A

5 Produces identical copies of the genetic information so that each daughter cell can be supplied with the information it needs to carry out all its activities and functions.

***Exam* tip:** Why not learn DNA replication as a flowchart detailing each stage? It will be very useful in answering extended response questions.

Transcription of DNA

Q1 Name the site of transcription of DNA.

Q2 Describe the process of transcription of DNA.

Q3 Give *three* structural differences between DNA and RNA molecules.

Q4 Give the term for the triplets of bases on mRNA, each of which codes for an amino acid.

Q5 Describe the role of mRNA molecules in the synthesis of protein.

ANSWERS

1 nucleus

2 DNA uncoils then unzips, bases on free RNA nucleotides pair with their complementary DNA bases, sugar–phosphate bonds form between the nucleotides of RNA to form a molecule of messenger RNA.

3 DNA is double stranded and RNA is single stranded; DNA contains thymine but in RNA this base is replaced by uracil; DNA nucleotides contain deoxyribose sugar but RNA nucleotides contain ribose sugar.

4 codons

5 mRNA carries the genetic code from the nucleus to the ribosomes.

***Exam* tip:** Questions asking you to compare DNA and RNA need a *full* comparison to be given full marks.

Translation of mRNA

Q1 Name the site of translation of mRNA.

Q2 Name the molecule that carries a specific amino acid to the ribosome.

Q3 What name is given to the triplet of bases on a tRNA molecule?

Q4 Give the term used to describe the relationship between codons of mRNA and matching triplets of bases on tRNA.

Q5 Describe the process of translation.

ANSWERS

1 ribosomes

2 transfer RNA (tRNA)

3 anticodon

4 complementary

5 A tRNA molecule carrying specific amino acids approaches an mRNA molecule at a ribosome. Anticodons of tRNA line up against their complementary codons on mRNA, and the amino acids carried by tRNA link up to form protein molecules.

***Exam* tip:** Learning the processes of transcription and translation as extended response answers is really good exam preparation.

Secretion of protein

Q1 Describe the role of the ribosomes.

Q2 Describe the distribution of ribosomes within cells.

Q3 Describe the role of the rough endoplasmic reticulum.

Q4 Describe the role of the Golgi apparatus.

Q5 Describe what happens to vesicles after being produced by the Golgi apparatus.

ANSWERS

1 The ribosomes are the sites of the translation of mRNA/sites of protein synthesis.

2 Most are found attached to the membranes of the rough endoplasmic reticulum, but some are free in the cytoplasm of the cells.

3 The rough ER transports proteins within the cell.

4 The Golgi apparatus processes and packages protein into vesicles for secretion.

5 The vesicles fuse with the plasma membrane, releasing their contents outside the cell.

***Exam* tip:** The content covered in this topic is associated with cells that synthesise and secrete proteins such as enzymes and hormones.

Structure of viruses

Q1 Name the substance that forms the coat of viruses.

Q2 Name the substance enclosed within the coat of a virus.

Q3 Describe the structure of a virus.

Q4 Give an indication of the relative size of virus particles.

Q5 Explain the nature of viruses and how they are able to survive.

ANSWERS

1 protein

2 nucleic acid/DNA/RNA

3 Nucleic acid molecules enclosed within protein coats.

4 Much smaller than cells.

5 They rely on host cells for their survival and reproduction. They are obligate parasites.

***Exam* tip:** All the information in this section could be included in an extended response question that asks about the *nature* of viruses.

Viral replication

Q1 Describe what happens to the metabolism of a host cell immediately following the introduction of viral nucleic acid.

Q2 Describe the first process through which viral nucleic acid passes within a host cell and name the substances supplied by the host cell to allow this process.

Q3 Viral nucleic acid is translated by the host cell's metabolism.

Describe the result of the translation of viral nucleic acid.

Q4 New viruses are assembled within host cells.

Describe this process.

Q5 Describe what happens to new viruses produced within a host cell.

ANSWERS

1 The virus alters the cell's metabolism, causing the production of more viruses.

2 Viral nucleic acid replicates using nucleotides, ATP and enzymes supplied by the host cell.

3 Viral nucleic acid is translated into protein, which will form the coats of more viruses.

4 New viruses are produced when viral nucleic acid is surrounded or enclosed by a viral protein coat.

5 Lysis *or* new viruses burst out of the host cell.

***Exam* tip:** Candidates often lose marks in this topic by using descriptions that don't show full understanding. For example, always refer to *viral* nucleic acid – host cells have nucleic acid too!

Antibody production

Q1 The presence of viruses in the human body triggers antibody production.

Name the term for any substance that triggers this response.

Q2 Name the blood cell that produces antibodies.

Q3 Describe the function of antibodies.

Q4 Explain why an antibody produced in response to one antigen is not effective against different antigens.

Q5 Describe how antigens are involved in tissue rejection following a transplant, and name the treatment given to prevent rejection.

ANSWERS

1 antigen

2 lymphocyte

3 Antibodies render antigens harmless.

4 Antibodies are specific to the antigen that triggered their production.

5 Transplanted tissue has antigens that trigger antibody production in the transplant recipient. These antibodies attack the transplant.

Treatment involves use of immuno-supressor drugs.

***Exam* tip:** Remember that the **antigens** cause **anti**bodies to be **gen**erated by lymphocytes. More difficult questions on this topic need knowledge of the *specificity* of antibodies.

Phagocytosis

Q1 Give the term used to describe the engulfing and digestion of foreign material by cells.

Q2 Name the type of blood cell involved in phagocytosis.

Q3 Describe how the plasma membrane is involved in phagocytosis.

Q4 Describe the importance of lysosomes in phagocytosis.

Q5 State why phagocytosis is described as non-specific.

ANSWERS

1 phagocytosis

2 phagocyte

3 The plasma membrane engulfs foreign material, enclosing it into a vacuole.

4 Lysosomes fuse with vacuoles and add the digestive enzymes they contain into the vacuole to destroy the contents.

5 Phagocytosis is a general process triggered by a huge variety of foreign materials.

***Exam* tip:** Make sure you know the role of the lysosomes in phagocytosis.

Remember that it is the phagocyte *cells* that surround and engulf the foreign particles, *not* the lysosomes.

Cellular defence in plants

Q1 Name a substance produced by some plants to isolate injured areas.

Q2 Explain why isolation of injured areas is an advantage in the survival of a damaged plant.

Q3 Name *three* chemicals produced by various plants to deter grazing by herbivores.

Q4 Give *two* ways in which chemicals can deter grazing by herbivores.

Q5 Explain the difference between deterring grazing and tolerating it.

ANSWERS

1 resin

2 It acts as a barrier, preventing any infection entering the injured area from spreading.

3 Tannins, nicotine, cyanide.

4 By poisoning the herbivore or by being unpalatable to herbivores.

5 Deterring stops or reduces the grazing. Tolerating indicates that recovery following grazing is possible.

***Exam* tip:** Remember Three Nasty Chemicals + Resin.

T – tannin, N – nicotine, C – cyanide.

Sexual reproduction and variation

Q1 State what is meant by the term sexual reproduction.

Q2 Give the importance of sexual reproduction in a population.

Q3 Explain the importance of sexual reproduction in long-term evolutionary change.

Q4 Name the type of cell division that leads to variation in the daughter cells.

Q5 Name *two* processes occurring during meiosis that lead to variation within the gametes produced.

ANSWERS

1 Reproduction involving the fertilisation of one gamete by another.

2 It maintains variation within a species.

3 It provides the variation upon which natural selection can act.

4 meiosis

5 Crossing over and independent assortment.

***Exam* tip:** Remember that crossing over and independent assortment both occur in the *first* meiotic division.

Meiosis

Q1 Give a location in a flowering plant and in a mammal where cells undergo meiosis.

Q2 Homologous chromosomes are the same size and shape.

State *one* other feature that homologous chromosomes have in common.

Q3 Apart from mutations, state *two* other events which occur during meiosis that result in genetic variation.

Q4 Name the cells that undergo meiosis and the cells that are produced by meiosis.

Q5 The events below occur during the process of meiosis:

A Chromatids are pulled apart.
B Homologous pairs of chromosomes separate.
C Chromosomes form homologous pairs.
D Chiasmata form and crossing over occurs.

Give the order in which they occur.

ANSWERS

1 Flowering plant: anthers *or* ovaries.
Animals: testes *or* ovaries.

2 They have the same genes in the same order.

3 Crossing over and independent assortment.

4 Gamete mother cells undergo meiosis and gametes are formed.

5 C, D, B, A

Exam tip: Meiosis occurs in *organs* such as testes and ovaries and produces *gametes* such as sperm and ova. Don't confuse the gametes with the organs in which they are produced!

Crossing over and independent assortment

Q1 Name the points of contact that can form between homologous chromosomes during meiosis.

Q2 Explain the importance of chiasmata formation.

Q3 Describe how recombination can lead to variation in the genotype of gametes.

Q4 Independent assortment of chromosomes during meiosis is a source of genetic variation.

Describe the behaviour of chromosomes during the first meiotic division that leads to independent assortment.

Q5 Explain how independent assortment can lead to the production of new phenotypes.

ANSWERS

41 Crossing over and independent assortment

1 chiasmata

2 Chiasmata lead to crossing over and so increase variation.

3 Recombination allows new combinations of alleles to be created in gametes.

4 Homologous chromosomes align randomly and when they pull apart the chromosomes can assort in different combinations.

5 Because gametes contain different combination of alleles and fertilisation is random, new phenotypes are produced in the new generation.

Exam tip: Watch the spelling of chiasma and chiasmata – you could lose marks for the wrong spelling here!

Dihybrid cross (1)

Q1 The actual phenotype ratio obtained following a genetic cross may differ from the expected ratios.

Suggest *two* reasons why the actual numbers of offspring observed may differ from the expected numbers in a dihybrid cross.

Q2 State what is meant by an allele.

Q3 Give the term that refers to a genotype in which the alleles of a particular gene are the same.

Q4 Give the term that refers to a genotype in which the alleles of a particular gene are different.

Q5 Explain what is meant by the terms dominant and recessive alleles.

ANSWERS

42 Dihybrid cross (1)

1 Fertilisation is random and numbers of offspring may be too small for expected ratios to be obtained.

2 An allele is a form of a gene.

3 homozygous

4 heterozygous

5 Dominant alleles always show in the phenotype of an organism. Recessive alleles only show in the phenotype when in the homozygous state.

***Exam* tip:** Homologous chromosomes have the same **genes** but *not* necessarily the same **alleles** – many candidates make this mistake.

Dihybrid cross (2)

These questions involve crosses in which the genes being investigated are located on different chromosomes.

One gene has alleles **G** and **g** and the other gene has alleles **R** and **r**.

Q1 Give the genotypes of the gametes from a parent with the genotype **GgRr**.

Q2 Give the phenotype ratio of the offspring expected from two parents that each have a heterozygous genotype (**GgRr** × **GgRr**).

Q3 Give the phenotype ratio of the offspring expected from a parent that is heterozygous and another parent that is homozygous recessive (**GgRr** × **ggrr**).

Q4 State the expected phenotype ratio from the cross **Ggrr** × **ggRr**.

Q5 State the expected genotype ratio from the cross **GGrr** × **ggRR**.

ANSWERS

1 GR, Gr, gR, gr

2 9:3:3:1

3 1:1:1:1

4 1:1:1:1

5 *All* offspring will have the genotype **GgRr**.

***Exam* tip:** When working out ratios in your exam, use the space for calculation that is provided. Markers look at working because there may be marks there. Also – the answer is *not* always 9:3:3:1!

Linkage and recombination

Q1 Give the term used to describe genes present on the same chromosome.

Q2 State the genotypes of the gametes that would be produced by a parent with a genotype of **GgRr**, assuming that the genes are linked.

Q3 State the significance to linked genes of crossing over during meiosis.

Q4 Explain why recombinant phenotypes may appear in offspring even when the genes involved are linked.

Q5 Explain the differences in crossover values or frequencies between linked genes.

ANSWERS

44 Linkage and recombination

1 linked

2 **GR** and **gr** *or* **Gr** and **gR**.

3 The linked genes may be separated and then able to recombine, giving increased variation in the gametes.

4 Crossing over and recombination have affected some of the gametes produced.

5 Genes that are close to each other on a chromosome are seldom separated by chiasmata and their crossover values are low. Genes that are further apart on a chromosome are more likely to be separated by chiasmata and their crossover values are high.

***Exam* tip:** Remember that you can be asked to map the relative positions of up to *four* genes on a chromosome given the percentage recombination frequencies. The lower the frequency the closer the genes.

Sex linkage

Q1 Explain why males are more likely to be affected by recessive sex-linked conditions than females.

Q2 Name the chromosome that carries sex-linked alleles.

Q3 Give *two* examples of sex-linked conditions in humans.

Q4 State the term given to a female who is heterozygous for a recessive sex-linked condition.

Q5 Give the ratio of phenotypes expected from a cross between a female heterozygous for a recessive sex-linked condition and an affected male.

ANSWERS

1 Males have only one X chromosome and so inherit only one copy of each sex-linked allele and so this will always show in their phenotype. Females inherit two copies of each sex-linked allele and both would have to be recessive for them to show the condition in their phenotype.

2 X chromosome

3 haemophilia, red–green colour deficiency

4 carrier

5 1 carrier female : 1 affected female ; 1 unaffected male ; 1 affected male

***Exam* tip:** When giving the symbols for sex-linked alleles you must give the sex chromosomes, and the allele symbols must be *superscripts*, for example X^RX^r.

Remember, the Y chromosome does not have sex-linked alleles.

Mutant alleles

Q1 Give a definition of a mutation.

Q2 State *two* characteristic features of mutant alleles.

Q3 Name *two* mutagenic agents.

Q4 Name *two* types of gene mutation that affect the number of bases in a gene.

Q5 Name *two* types of gene mutation that affect base sequence *without* affecting base number in a gene.

ANSWERS

1 Changes in the genetic information of an organism that can affect individual genes or whole chromosomes.

2 Random occurrence and low frequency.

3 Chemicals, e.g. colchicine, mustard gas, *and* irradiation by, e.g., X-rays, UV light.

4 deletion, insertion

5 substitution, inversion

***Exam* tip:** For mutations – remember Rolf!

Random occurrence at low frequency.

Gene mutations

Q1 Explain how a gene mutation could lead to:

a the production of an abnormal cell membrane protein
b a block in a metabolic pathway.

Q2 Describe the effect of an insertion or a deletion mutation on amino acid sequences in the protein molecule coded for.

Q3 Describe the effect of a substitution or an inversion mutation on amino acid sequences in the protein molecule coded for.

Q4 Give *two* examples of gene mutations that cause only one or two amino acids in the protein coded for to be changed.

Q5 Give *two* examples of gene mutations that cause very significant changes in the sequence of amino acids in the protein coded for.

ANSWERS

1 a A change in the base order of DNA would result in changes to the order of amino acids in the cell membrane protein coded for, producing an abnormal protein.

b A change in the base order of DNA would result in changes to the order of amino acids in the enzyme coded for. The enzyme would not function and so a block in the metabolic pathway controlled by the enzyme would occur.

2 There would be a major effect on the protein coded for. All of the amino acids in the sequence following the mutation would be affected.

3 There would be a minor effect on the protein coded for. Only one or two amino acids in the sequence would be affected.

4 substitution and inversion

5 deletion and insertion

***Exam* tip:** To remember the gene mutations – try **DIGSI**:

Deletion and **I**nsertion (**G**ene mutation) **S**ubstitution and **I**nversion.

Chromosome mutations

Q1 Name *four* types of chromosome mutation.

Q2 Describe what happens in each of the chromosome mutations listed:

a translocation
b duplication
c deletion
d inversion.

Q3 Name the type of chromosome mutation that can result in a change to the chromosome number in gametes.

Q4 Give an example of a human condition that results from a change in the number of chromosomes in a fertilized egg.

Q5 Explain why plants with one or more sets of chromosomes greater than the normal diploid number might display increased vigour.

ANSWERS

1 translocation, duplication, deletion, inversion

2 a Translocation – a section of one chromosome detaches and becomes attached to a different chromosome.

b Duplication – a section of chromosome is copied so that the sequence of genes it carries appears twice.

c Deletion – a section of chromosome is detached and is lost permanently.

d Inversion – a section of chromosome detaches and rotates through 180°.

3 non-disjunction

4 Down's syndrome (Trisomy 21)

5 The plants benefit from the dominant genes from both parents, which contribute to the improved growth and resistance to disease, frost, drought, etc.

***Exam* tip:** Note that the terms deletion and inversion are also used to describe certain *gene* mutations – be careful not to confuse these.

Polyploidy

Q1 State what is meant by the term polyploidy.

Q2 Explain how polyploidy can arise.

Q3 Explain why hybrid plants are often sterile.

Q4 Explain what is meant by the term hybrid vigour.

Q5 Give *one* advantage to humans of polyploidy in crop plants such as wheat.

ANSWERS

1 Possession of one or more sets of chromosomes above the normal diploid number.

2 Complete non-disjunction of chromosomes during meiosis can give rise to diploid gametes. If a diploid gamete was fertilised by a normal gamete, the resulting zygote would be polyploid with three sets of chromosomes.

3 Hybrid plants usually have no homologous chromosomes and so pairing during meiosis cannot occur and no gametes are produced.

4 Hybrid plants usually have greater yield, grow faster and are often resistant to disease, frost or drought.

5 Modern wheat plants show hybrid vigour and have increased yields and resistance to disease.

***Exam* tip:** Remember that polyploidy in crop plants can be an advantage to growers because it often confers increased *vigour* to crops – they grow faster, stronger and bigger and are more resistant to disease and poor growing conditions.

Concept of species

Q1 Give the definition of a species in terms of fertility.

Q2 Give the definition of a species by reference to the gene pool.

Q3 Explain what is meant by the evolution of species.

Q4 State the meaning of the term extinction.

Q5 Give the meaning of the phrase 'survival of the fittest'.

ANSWERS

1 Members of a species are able to interbreed to produce fertile young.

2 A species has a common gene pool, because its members can freely interbreed with each other.

3 Evolution refers to gradual changes in species, which eventually give rise to new species.

4 Extinction refers to the complete failure of any members of a population to survive.

5 Members of a species with favourable characteristics that give them a selective advantage, survive to breed.

***Exam* tip:** Many candidates forget that, while certain different species may occasionally interbreed with each other, they *do not* produce fertile young.

Natural selection

Q1 Describe what is meant by the term natural selection.

Q2 Describe what is meant by selection pressure.

Q3 Name the selection pressure that removes individuals from a population of peppered moths in which some moths are speckled and some are melanic, depending on their level of camouflage.

Q4 Explain how the presence of an antibiotic in the environment of a species of disease-causing bacterium can lead to the development of a resistant strain.

Q5 Give *two* examples of high-speed evolution.

ANSWERS

51 Natural selection

1 The survival of organisms best adapted to their environment.

2 Factors such as predation or disease, which can remove organisms from a population.

3 predation

4 An antibiotic is a selection pressure that removes all but resistant bacteria from a population. The resistant individuals survive, reproduce and pass their resistance on to offspring, producing a resistant strain.

5 Evolution of melanism in peppered moths and antibiotic resistance in bacteria.

***Exam* tip:** Although there are some high-speed examples, evolution usually occurs *very slowly*, taking many millions of generations to produce new species. In writing about evolution, candidates should normally emphasise its *very slow* nature.

Isolation and speciation

Q1 Name *three* isolating mechanisms that can be involved in speciation.

Q2 State the importance of isolating mechanisms in the evolution of new species.

Q3 Explain how mutations and natural selection account for the different gene pools between two sub-populations that have been separated by an isolating mechanism for a long time.

Q4 Give the meaning of the term speciation.

Q5 Describe the evidence which would confirm that two sub-populations of the same species had evolved to become two different species.

ANSWERS

1 Geographical, ecological and reproductive.

2 They act as barriers to gene exchange.

3 Mutation affects the gene pools of isolated sub-populations differently. Selection pressures act differently on each sub-population. The changes to the gene pool of one sub-population cannot be passed to other sub-populations because of the isolating barrier.

4 The formation of two or more species from one original species.

5 They would be unable to interbreed to produce fertile offspring.

***Exam* tip:** Remember **GERM**.

Geographical, **E**cological and **R**eproductive isolation **M**echanisms.

Adaptive radiation

Q1 Explain what is meant by the term adaptive radiation.

Q2 Explain what is meant by a common ancestor.

Q3 Explain how adaptive radiation allows the survival of a group of similar species.

Q4 Give a piece of evidence which would confirm that a group of similar species had arisen by adaptive radiation.

Q5 Describe how the differences in the structure of their beaks allow the different species of Darwin's finches to survive together on the Galapagos Islands.

ANSWERS

53 Adaptive radiation

1 Evolution of a group of new species from a common ancestor. The species differ from each other in an important adaptation such as their feeding structures.

2 An original species that has speciated to give a group of similar species.

3 Each species in the group has evolved different characteristics, which reduces inter-specific competition.

4 Evidence of a common ancestor.

5 Their beaks differ in size and shape to exploit different foods and so reduce inter-specific competition.

***Exam* tip:** When describing adaptive radiation, ensure that you say what the adapted structures are designed for – the *different* beaks of Darwin's finches are adapted for *different* foods.

Conservation

Q1 Give a reason for the conservation of species.

Q2 Explain how the creation of a wildlife reserve might conserve a species.

Q3 State what is meant by captive breeding.

Q4 Give an example of a cell bank and explain its contribution to conservation.

Q5 State why the maintenance of genetic diversity is important in the conservation of species.

ANSWERS

54 Conservation

1 Maintenance of biodiversity.

2 Allows measures such as the prevention of hunting to be implemented *or* allows maintenance of a specific habitat *or* increases the breeding potential of a species.

3 The breeding of endangered species in zoos or rare breed farms for release back into the wild.

4 Sperm bank – allows artificial insemination of females of the species to produce young, which can be released into the wild.

5 Genetic diversity is essential to species because it increases the chance that some individuals survive changing selection pressures.

***Exam* tip:** It is probably safer to use the methods of conservation described in the Arrangements for Higher Biology – *wildlife reserves, captive breeding and cell banks* – although these are not the only methods!

Selective breeding and hybridisation

Q1 Describe the process of selective breeding.

Q2 Give *one* specific example in which artificial selection has been used to human advantage.

Q3 Explain why selective breeding has limitations.

Q4 Explain what is meant by the term hybridisation.

Q5 Explain why hybridisation is limited.

ANSWERS

1 Individuals with desirable characteristics are chosen to breed together to produce offspring which show these characteristics.

2 Many possible examples – increased beef yield or milk production in cattle *or* better wool quality in sheep *or* greater speed in racehorses *or* increased yield in crops, etc.

3 It is limited by the existing variation of the species.

4 Crossing two closely related species to obtain hybrid vigour in the offspring or to obtain offspring showing characteristics of both parents.

5 Offspring may not be produced or are infertile.

***Exam* tip:** When giving an example of artificial selection, make sure you mention the *characteristic* that has been improved, for example improved beef yield in cattle.

Genetic engineering

Q1 Name a technique that could be used to locate the position of a gene on a chromosome.

Q2 Describe *two* procedures used in genetic engineering that use endonuclease enzymes.

Q3 Describe the use of ligase enzymes in genetic engineering.

Q4 Name *one* human hormone that is produced by genetically engineered bacteria.

Q5 Give *one* advantage of the use of genetic engineering in the production of substances such as insulin.

ANSWERS

1 Gene probes or characteristic banding patterns on stained chromosomes.

2 Cutting genes from a chromosome of the donor organism *and* opening a plasmid of the recipient bacterium.

3 To seal the genes of one organism into the genome of another, e.g. into the plasmid of a bacterium.

4 Insulin *or* human growth hormone (GH).

5 Product is pure *or* can be produced quickly *or* can be produced in large quantities.

***Exam* tip:** The examples of genetic engineering that you *must* know about are the manufacture of insulin *and* human growth hormone.

Somatic fusion

Q1 Give the meaning of the term somatic fusion and describe why somatic fusion is carried out.

Q2 State the problem in plant breeding that is overcome by somatic fusion.

Q3 Name the enzyme used in somatic fusion to remove cell walls.

Q4 Give the term used to describe a cell formed after its cell wall is removed.

Q5 Describe the stage following the production of protoplasts in somatic fusion.

ANSWERS

1 Fusion of somatic cells from two different plant species to produce a new species that has the desirable characteristics of both parents.

2 Sexual incompatibility between different plant species.

3 cellulase

4 protoplast

5 The encouragement of the fusion of the protoplasts.

***Exam** tip:* It is advisable to use the term *sexual incompatibility* to describe why some plant species are crossed using somatic cell fusion, rather than trying to express this in other words.

Osmoregulation in fish

Q1 Name the process causing the gain or loss of water from the gill cells of a fish.

Q2 Give *two* adaptations for osmoregulation in freshwater fish.

Q3 Give *two* adaptations for osmoregulation in saltwater bony fish.

Q4 State the change in the glomerular filtration rate when salmon or eels migrate from fresh water to the sea.

Q5 Describe the role of chloride secretory cells of:

a salmon in sea water
b eels in fresh water.

ANSWERS

58 Osmoregulation in fish

1 osmosis

2 Many large glomeruli *or* high filtration rate of kidneys *or* chloride secretory cells in gills actively absorb salts (any two).

3 Few small glomeruli *or* low filtration rate of kidneys *or* chloride secretory cells in gills actively excrete salt (any two).

4 Filtration rate decreases.

5 **a** They actively excrete salts.

b They actively absorb salts.

***Exam* tip:** Make sure that you use the term *chloride secretory cells* in the *gills* of fish – secretory cells alone won't do!

Water conservation in desert mammals

Q1 Desert rats have nocturnal habits.

Explain the survival value of this behavioural adaptation.

Q2 Desert rats burrow in sandy soil.

Explain the survival value of this behavioural adaptation.

Q3 Give *two* physiological adaptations of the kidneys of the desert rat to reduce water loss.

Q4 Describe how the large intestine of the desert rat is adapted to reduce water loss.

Q5 Explain why the mouth and nasal passages of the desert rat have relatively dry surfaces.

ANSWERS

59 Water conservation in desert mammals

1 The coolness of night reduces evaporation of water from their bodies *or* their failure to sweat does not produce overheating.

2 Burrowing provides a moist and cool environment during the day which reduces water loss by evaporation.

3 Low glomerular filtration rate *and* high levels of water reabsorption by kidney tubules.

4 It allows efficient reabsorption of water.

5 Prevent excessive evaporation of water during breathing.

***Exam* tip:** You must know the difference between adaptations of the desert rat that are *physiological* and those that are *behavioural*.

Transpiration (1)

Q1 Give definitions of the following:

- **a** transpiration
- **b** the transpiration stream.

Q2 Describe the process by which water moves into a root hair cell from the surrounding soil.

Q3 Describe the movement of water from a root hair cell across the root cortex to the xylem.

Q4 Explain how the forces of adhesion and cohesion help in the maintenance of the transpiration stream.

Q5 Describe how water passes from the xylem vessels in a leaf to the atmosphere.

ANSWERS

1 a Evaporation of water from leaf surfaces through the stomata.
  b Continuous columns of moving water from roots, through stems to leaves.

2 By osmosis, from a region of high water concentration in the soil to a region of lower water concentration in the root hair cell.

3 Water moves by osmosis across the cortex cells from high to lower water concentration.

4 Cohesion holds water molecules together and adhesion holds the water molecules to the walls of the xylem to maintain a continuous column of water through the plant.

5 Water evaporates into leaf air spaces then diffuses through the stomata.

***Exam* tip:** Never say *along* a concentration gradient in describing osmosis – it has to be *down*!

Transpiration (2)

Q1 State *three* changes to environmental conditions that could result in a *decrease* in the rate of transpiration.

Q2 Explain the change in rate of transpiration that would be expected if the temperature in the environment of a plant *increased*.

Q3 Explain the change in the rate of transpiration that would be expected if there was an *increase* in the air movements surrounding the leaves of a plant.

Q4 The transpiration stream transports water to plant cells for photosynthesis.

Give *one* other role of the transpiration stream in transporting materials.

Q5 Other than its role in transport, give *two* other benefits to plants of the transpiration stream.

ANSWERS

1 Decrease in temperature *or* decrease in air movements *or* decrease in light intensity *or* increase in humidity (any three).

2 Increase in temperature would increase the rate of evaporation, so increasing the transpiration rate.

3 Increase in air movements would increase the rate of evaporation by replacing moist air with drier air, so increasing the transpiration rate.

4 Transport of minerals.

5 Cooling *and* maintaining turgor.

***Exam* tip:** It is worth learning a short list of factors that affect the rate of transpiration – but remember to say whether they increase or decrease the rate.

Stomata

Q1 State the role of guard cells.

Q2 State the changes in turgidity that would cause the opening **and** closing of stomata.

Q3 Explain how changes in the turgor of guard cells can affect the rate of transpiration.

Q4 Stomata open as conditions change from dark to light.

Explain why this is important to the survival of plants.

Q5 State the advantage to plants in having their stomata closed in dark conditions.

ANSWERS

1 Guard cells change shape to control the opening and closing of the stomata.

2 Stomata open when guard cells are turgid *and* close when guard cells are flaccid.

3 Increased turgor opens stomata allowing water vapour to escape, so increasing the rate of transpiration.

4 Open stomata are needed to allow the absorption of carbon dioxide for photosynthesis during light hours.

5 Conservation of water.

***Exam* tip**: You only need to know that **open** stomata have *turgid* guard cells and **closed** ones have *flaccid* guard cells. Details are not needed.

Hydrophytes

Q1 State what is meant by the term hydrophyte.

Q2 Explain why some hydrophytes have flexible stems.

Q3 Give *one* structural adaptation of the stems of some hydrophytes that gives the stems increased flexibility.

Q4 Explain why some hydrophytes have many large air spaces in their leaves.

Q5 Explain why hydrophytes only have stomata on the upper surfaces of their leaves.

ANSWERS

1 Plants that live totally or partially submerged in water.

2 To avoid breaking in flowing water *or* to allow response to change in water level.

3 They have central xylem vessels *or* reduced xylem vessels.

4 Makes them buoyant, so keeping them in the light for photosynthesis and in the atmosphere for gas exchange.

5 To allow exchange of gases with the atmosphere.

Exam tip: When talking about hydrophytes floating, remember to say why they need to float and mention light for photosynthesis or stomata in the atmosphere for gas exchange.

Xerophytes

Q1 State what is meant by the term xerophyte.

Q2 Explain how the following adaptations reduce water loss from leaves:

a presence of hairs on a leaf surface
b few small leaves.

Q3 Explain how sunken stomata are an advantage in hot, dry and windy conditions.

Q4 Explain the advantage to a xerophyte in having reversed stomatal rhythm in which stomata are open at night rather than during the day.

Q5 Describe *two* adaptations of the root systems of plants that grow in soils with a low water availability, and explain how each assists the uptake of available water.

ANSWERS

1 Plants that are adapted for survival in habitats with low water availability *or* where conditions produce very high rates of transpiration.

2 a They reduce air movements and trap moist air, lowering the transpiration rate.

b It reduces the surface area for evaporation.

3 Reduce air movements and trap moist air around the stomata, lowering transpiration rates.

4 Reduces transpiration during hot parts of the day.

5 Deep roots – reach water from the low water table in the soil.

Wide-spreading shallow roots – absorb any rain water before it can evaporate or drain away.

***Exam* tip:** Since you need to *explain* how the adaptations work, it's better to know a couple of examples thoroughly!

Foraging and competition

Q1 Animals must forage economically.

Explain what is meant by this statement in terms of energy gain and loss.

Q2 Following a successful hunt, wild dogs may be displaced from their kill by spotted hyenas.

State the type of competition that this shows.

Q3 Define the term intraspecific competition.

Q4 Explain why intraspecific competition is more intense than interspecific competition.

Q5 Give *two* environmental factors for which plants compete.

ANSWERS

1 Energy gained in food must exceed that spent in foraging *or* there must be a net gain of energy.

2 interspecific

3 Competition within one species.

4 Members of the species compete for almost identical resources and have similar adaptations.

5 Light, space, water, minerals (any two).

***Exam* tip:** Make sure you understand the idea of *net* gain of energy. The energy gained in food must be greater than the energy used in foraging.

Dominance hierarchy, co-operative hunting and territorial behaviour

Q1 Many mammals that form social groups demonstrate dominance hierarchies.

Define the term dominance hierarchy.

Q2 Give *one* advantage of a dominance hierarchy in a social organisation.

Q3 Describe what is meant by the following terms:

a co-operative hunting
b territorial behaviour.

Q4 Give *one* advantage of co-operative hunting to a social group of predatory animals.

Q5 Explain the advantage of territorial behaviour.

ANSWERS

1 A system involving a rank order of individuals, in which dominant individuals have priority over subordinate ones in feeding, mating, etc.

2 Prevents the waste of energy in intraspecific competition *or* ensures that individuals with the most favourable characteristics survive to breed.

3 **a** hunting together as a social group to increase efficiency and success

b the marking and defence of an area of habitat for exclusive use.

4 Allows larger prey to be caught *or* increases the percentage of successful hunts *or* reduces the energy used per individual.

5 Reduces competition.

Exam **tip**: You must be clear about the difference between interspecific and intraspecific competition.

Plants and food

Q1 Explain why there could be an increase in plant species diversity in some habitats as a result of a moderate increase in grazing intensity by herbivores.

Q2 Explain why very high grazing intensity may decrease plant species diversity in some habitats.

Q3 State the meaning of the term compensation point.

Q4 Describe the difference in compensation point of a sun plant and a shade plant.

Q5 Explain why shade plants can survive in habitats with low light intensities.

ANSWERS

67 Plants and food

1 Grazing holds dominant or more vigorous grasses in check, allowing the survival of less dominant and vigorous species.

2 Some plants are eaten to extinction by the herbivores.

3 The level of light intensity at which the gain of food or glucose by photosynthesis exactly balances the loss by respiration *or* the rate of photosynthesis equals the rate of respiration *or* the rate of CO_2 uptake equals the rate of CO_2 output.

4 Shade plants reach their compensation point at lower light intensities than sun plants.

5 Some shade plants have a wide range of photosynthetic pigments that allow them to use light which has been transmitted through the sun plants above them.

***Exam* tip:** When describing the adaptations of shade plants, it is useful to mention that they can usually reach their compensation point at *lower* light intensities than sun plants.

Avoidance and habituation

Q1 Give *one* example of an escape response by an animal.

Q2 If disturbed while feeding, a snail withdraws into its shell as an escape response.

Name this type of behaviour.

Q3 Name the type of behaviour in which an animal learns not to respond to repeated harmless stimuli.

Q4 Explain the advantage of habituation.

Q5 Explain why it is important that habituation is short term in its effect.

ANSWERS

1 Examples include withdrawal reflexes in various invertebrates such as snails or worms.

2 avoidance behaviour

3 habituation

4 The animal saves energy by no longer responding, and habituation allows gain of energy in continued foraging.

5 New stimuli may be dangerous and require avoidance behaviour.

***Exam* tip:** It is essential to emphasise that habituation is an adaptation that saves the animal *energy*.

Individual and social defence

Q1 Explain how living in social groups may help animals in defence against predators.

Q2 Explain how collecting into large swirling flocks may protect starlings from predators.

Q3 Explain how feeding in large flocks may protect certain species of wading birds from predators.

Q4 Individual animals may have coloration that protects them against predators.

Explain the survival value of each of the following coloration adaptations:

a camouflage in the peppered moths
b warning coloration in stinging wasps.

Q5 Give *one* example of an animal that has structural adaptations for individual defence.

ANSWERS

1 The animals can co-operate to repel a predator *or* can alert each other to predation danger *or* can confuse a predator by movements.

2 It serves to confuse the predator, making survival of an individual more likely.

3 The birds can take turns in watching for predators, so allows more feeding time per individual.

4 **a** Helps to prevent being spotted by predators.

b Alerts a predator to avoid the individual by linking colour with stinging potential.

5 Snail has a shell *or* hedgehog has spines, etc.

***Exam* tip:** Learn a few examples of both individual and social mechanisms for defence – ensure that you can explain why the adaptations work.

Defence in plants

Q1 Explain why having deep roots allows a plant to tolerate grazing by herbivores.

Q2 Explain why having meristems at the base of their leaves allows grass species to tolerate grazing by herbivores.

Q3 Apart from deep roots and low meristems, give *one* other adaptation in plants that allows them to tolerate grazing by herbivores.

Q4 Nettles have stings on their leaves.

Describe the advantage to the species of this adaptation.

Q5 Apart from stings, give *two* other examples of structural defence mechanisms in plants.

ANSWERS

1 Plants can regenerate from the deep roots.

2 Plants can regenerate from the meristems that are too low to be grazed.

3 underground stems

4 It discourages grazing by herbivores.

5 thorns and spines

***Exam* tip:** Many candidates confuse the idea of *toleration* of grazing with the idea of *avoiding* grazing – tolerant plants grow back quickly, plants that avoid grazing discourage the grazer from eating at all!

Growth patterns

Q1 Explain the step-shaped growth pattern of a locust.

Q2 Describe the expected growth pattern for an annual plant as it grows from a seed through its year of growth.

Q3 Describe the expected growth pattern for a human from birth until age 20 years.

Q4 Explain why the measurement of average shoot length alone may not provide a reliable estimate of plant growth.

Q5 Explain why dry mass is a more reliable measurement of growth than the measurement of shoot length in plants.

ANSWERS

1 The bottom of each 'step' shows the beginning of moult when the locust sheds its exoskeleton. This is followed by a rapid expansion in length before the new exoskeleton hardens. No further change in length can occur before the next moult.

2 A slight drop in mass due to respiration is expected as the seed germinates. This is followed by a steady increase in mass over the growing season as the plant photosynthesises. Later, a decrease in mass is expected as the plant disperses its seeds and eventually dies.

3 Newborn babies decrease in mass for a short time before showing steady increase. There is a growth spurt during infanthood and another at puberty. A levelling out of growth would be expected by the age of 20.

4 Shoot length does not reflect the increase of mass of shoots and does not take into account growth of other regions of the plants such as the root system or lateral thickening.

5 Dry mass removes the variation in mass due to the variation in water content of plants, which does not reflect actual growth.

***Exam* tip:** These questions are usually poorly answered in the exam – learn these answers carefully.

Don't use the terms shell or outer coat to describe the locust's exoskeleton.

Meristems

Q1 State the meaning of the term meristem.

Q2 Describe the location of the meristems in an annual plant.

Q3 Name the meristematic tissue responsible for the formation of annual rings.

Q4 State the role of cambium in the formation of annual rings.

Q5 Explain the absence of meristems in animals.

ANSWERS

1 A meristem is the name given to a region in a plant where mitosis occurs.

2 Meristems are found at the tips of the roots and shoots, in buds and in cambium in the vascular bundles.

3 cambium

4 Cambium provides new cells, which can then differentiate into xylem or phloem tissue.

5 Animals grow over most of their bodies and so cell division is not restricted to precise regions.

***Exam* tip:** Answer 4 is seldom given by candidates – try to learn the wording given here.

Annual rings

Q1 Describe the appearance of spring and summer xylem vessels.

Q2 Describe *two* structural changes that take place as a cell becomes differentiated to form part of a xylem vessel.

Q3 What name is given to the area of a woody twig section that contains the xylem tissue growth occurring in a single year?

Q4 Explain how an examination of the annual rings of a cut woody stem could help identify the season in which it had been cut.

Q5 Explain how an examination of the annual rings of a cut woody stem could provide information about the growing conditions in a particular year.

ANSWERS

1 Spring vessels are wider and have fairly thin walls, but summer vessels are narrower with thicker walls, depending on growing conditions.

2 Cells lose their contents and their walls become inlaid with lignin during differentiation into xylem vessels.

3 An annual ring.

4 The outermost ring of xylem in a cut stem *or* the cells closest to the cambium were produced in the season in which the stem was cut. Wider xylem cells indicate spring and narrower cells indicate summer

5 Rings vary in their overall width depending on growing conditions, with narrow rings being characteristic of dry conditions or poor conditions for photosynthesis. Narrow rings could also indicate leaf damage due to disease, greenfly or caterpillars.

***Exam* tip:** Ensure that you emphasise that spring xylem vessels are *wider* – bigger will not do!

Jacob–Monod hypothesis

Q1 In the control of lactose metabolism in the bacterium *Escherichia coli*, identify the following:

- **a** the gene that produces the repressor molecule
- **b** the gene that produces lactose-digesting enzyme
- **c** the substance that acts as the inducer.

Q2 In the control of lactose metabolism in the bacterium *E. coli*, identify the part of the *E. coli* chromosome to which the repressor molecule binds in the absence of lactose.

Q3 In the Jacob–Monod hypothesis of gene action in *E. coli*, describe the sequence of events that would occur in the absence of the inducer.

Q4 In the Jacob–Monod hypothesis of gene action in *E. coli*, describe the sequence of events that would occur in the presence of the inducer.

Q5 Explain why it is an advantage to *E. coli* to control gene action.

ANSWERS

1 **a** the regulator gene
  b the structural gene
  c lactose

2 the operator

3 The repressor molecule binds to the operator which switches off the structural gene so that no enzyme is made.

4 The repressor molecule binds to the inducer leaving the operator free to switch on the structural gene and so allow synthesis of the enzyme.

5 Conserves energy and resources.

***Exam* tip:** Full understanding and use of correct terminology is essential.

Remember ROSE – Regulator, Operator, Structural, Enzyme.

Genes in metabolism

Q1 Explain the formation of different tissues during development, in terms of gene activity.

Q2 Explain how a gene mutation can give rise to conditions such as PKU.

Q3 Name the amino acid that cannot be metabolised in the normal way in those affected by PKU.

Q4 Name the amino acid that cannot be formed in the normal way from phenylalanine in those affected by PKU.

Q5 Explain how the symptoms of PKU arise.

ANSWERS

1 Different tissues develop due to the cells switching on certain genes and switching off others.

2 The mutated gene causes the absence of an enzyme *or* the production of an enzyme that does not work properly. This means that the usual substrate of the enzyme builds up and that its usual product is not produced.

3 phenylalanine

4 tyrosine

5 In PKU, the substrate builds up to the extent that it is converted to a toxic substance that affects the development of the nervous system, producing symptoms.

***Exam* tip**: Avoid saying that genes are switched on *and* off during differentiation! *Some* are off and *others* are on.

Pituitary hormones

Q1 Describe the role of human growth hormone.

Q2 Give the functions of TSH and thyroxine.

Q3 Name the gland responsible for the production and release of growth hormone and TSH.

Q4 Name the hormone that stimulates mitosis and protein synthesis in bones and muscle.

Q5 Name the hormone released by the pituitary gland that causes the thyroid gland to release thyroxine.

ANSWERS

1 Growth hormone stimulates the growth of bones and muscle by increasing the rate of amino acid uptake and protein synthesis in cells and by stimulating mitosis.

2 TSH regulates the activity of the thyroid gland, and thyroxine levels control the metabolic rate.

3 pituitary gland

4 growth hormone

5 TSH

***Exam* tip:** The role of growth hormone is poorly described by many candidates so learn Answer 1.

Plants and IAA (1)

Q1 State the location of IAA production in plants.

Q2 Describe the effect of IAA on cells in the shoot tip.

Q3 Name the growth response made by plant shoots to unidirectional light.

Q4 If a shoot is illuminated from one side, IAA is transported to the dark side.

Describe how the growth of the shoot would be affected by this treatment.

Q5 Pieces of shoot cut from behind the apices were immersed in IAA solution.

Describe the effect that would be expected on the growth of the pieces.

ANSWERS

1 In meristems at the shoot tips and root tips.

2 Promotes mitosis, elongation and differentiation.

3 phototropism

4 On the dark side, IAA stimulates elongation of cells. This makes the growth of the shoot uneven and causes it to grow towards the light source.

5 IAA causes cell elongation, so the pieces would be expected to lengthen.

***Exam* tip:** To remember the cellular effect of IAA, try **MED C** – **M**itosis, **E**longation and **D**ifferentiation, **C**ell level.

Plants and IAA (2)

Q1 State the meaning of the term apical dominance.

Q2 Describe the role of indole acetic acid (IAA) in apical dominance.

Q3 Describe the role of IAA in leaf abscission.

Q4 Explain the role of IAA in fruit formation.

Q5 Give *two* practical applications of IAA in horticultural practice.

ANSWERS

1 Growth of the apical bud and inhibition of the lateral buds.

2 The apical bud produces high concentrations of IAA, which inhibits the growth of lateral buds while promoting the growth of the apical bud.

3 A decrease in IAA concentration results in the formation of an abscission layer in the leaf stalk, causing it to break and the leaf to fall.

4 IAA from the seed causes the ovary wall to form a fruit.

5 herbicides and rooting powders

Exam tip: Remember the effect of IAA at organ level.

OLAF – Organ level, Leaf abscission, Apical dominance, Fruit formation.

Plants and GA

Q1 Name the substance made by the embryo of the barley grain that induces α-amylase production.

Q2 Name the layer in a barley grain that synthesises α-amylase in response to GA.

Q3 Explain the role of α-amylase in the process of germination.

Q4 Describe the effect of GA on internode cells.

Q5 Apart from its role in the germination of barley grains and in the elongation of the internode cells, give one other effect of GA.

79 Plants and GA

1 gibberellic acid (GA)

2 aleurone layer

3 α-amylase is responsible for the breakdown of starch to maltose, which can then be used by the seed to provide energy for germination.

4 GA causes the elongation of internode cells.

5 GA breaks the dormancy of buds in spring.

***Exam** tip*: Remember **GA-AID. G**ibberellic **A**cid – **A**mylase, **I**nternode cells, **D**ormancy.

Remember **SAM** – **S**tarch, **A**mylase, **M**altose.

Macro-elements and plant growth

Q1 Explain the importance of nitrogen to plants.

Q2 Explain the importance of magnesium to plants.

Q3 Explain the importance of phosphorus to plants.

Q4 Describe the importance of potassium to plant growth, and the symptoms of its deficiency.

Q5 Give *two* elements which, when deficient, can cause each of the following symptoms:

a red leaf bases
b chlorosis.

ANSWERS

80 Macro-elements and plant growth

1 Nitrogen is used in the synthesis of amino acids, proteins, nucleic acids, ATP.

2 Magnesium is used in the synthesis of chlorophyll, which allows plants to photosynthesise.

3 Phosphorus is used in the synthesis of nucleic acids and ATP.

4 Potassium is important in protein synthesis and membrane transport. Deficiency causes premature death of leaves and reduced growth.

5 **a** nitrogen, phosphorus

b nitrogen, magnesium

***Exam* tip:** Remember **N**itrogen for ami**N**o acid synthesis and protei**N**.

Lack of phospho**R**us causes **R**ed leaf bases and potassiu**M** is needed for **M**embrane transport.

Leaves become chlorotic when they are **M**i**N**us **M**agnesium or **N**itrogen.

Environmental influences in animals (1)

Q1 Which term is used to describe the effect on growth and development in many animals as a result of a change in the number of hours of light in a day?

Q2 Give *one* way in which a change in photoperiod affects behaviour in birds.

Q3 Mammals show photoperiodism.

Describe how one type of mammal behaviour can be affected by photoperiod.

Q4 Breeding behaviour in red deer starts in autumn.

Describe the environmental influence that triggers the start of breeding at this time of year.

Q5 Breeding behaviour in blue tits starts in spring.

Describe the environmental influence that triggers the start of breeding at this time of year.

ANSWERS

1 photoperiodism

2 Increasing photoperiod triggers the onset of courtship and breeding in birds.

3 Increasing photoperiod can bring the onset of breeding in small mammals. Decreasing photoperiod can bring the onset of breeding in larger mammals.

4 decreasing photoperiod

5 increasing photoperiod

Exam tip: *Do not* use the term day length when discussing the photoperiod, since the length of a day is always 24 hours and this includes hours of light *and* dark.

Environmental influences in animals (2)

Q1 Explain the importance of the following minerals in the growth and development of humans:

- **a** iron
- **b** calcium.

Q2 Describe the effect of nicotine on growth and development of a human fetus.

Q3 Explain the role of vitamin D and calcium in the prevention of rickets.

Q4 Describe the effect of alcohol on the growth and development of a human fetus.

Q5 Thalidomide is a drug prescribed for various conditions in humans.

Describe the effect of thalidomide on the growth and development of a human fetus.

ANSWERS

82 Environmental influences in animals (2)

1 a Iron is needed for haemoglobin, which is responsible for the transport of oxygen around the body, for many enzymes and hydrogen-carrying systems such as the cytochrome system.

  b Calcium is an essential component of bones and teeth and is needed for blood clotting.

2 Nicotine retards growth and mental development.

3 Vitamin D is required for the absorption of calcium from the small intestine, which is needed for the strengthening of bones.

4 Alcohol retards growth and mental development. It can produce facial abnormalities and heart defects in babies.

5 It causes limb deformities.

***Exam* tip:** Remember **C-BBC** which stands for **C**alcium needed for **B**ones and **B**lood **C**lotting.

Pregnancy **ANT**i drug campaign for the harmful effects of **A**lcohol, **N**icotine and **T**halidomide.

Effects of light on plant growth

Q1 A change in the photoperiod can initiate the flowering process.

State the meaning of the term photoperiod.

Q2 Describe the environmental condition that would result in phototropism.

Q3 Describe the environmental conditions in which the growing stem of a plant could become etiolated.

Q4 Describe *one* way in which an etiolated plant differs in appearance from a normal plant.

Q5 Potatoes are long-day plants.

In terms of changing photoperiod, describe what is meant by a long-day plant.

ANSWERS

1 The number of hours of light in a 24-hour period.

2 Light from one direction shining on a plant shoot.

3 darkness

4 Etiolated plants have small leaves, chlorosis, long internodes and long thin stems.

5 Long-day plants flower when the photoperiod is above a critical level *or* when the number of hours of darkness are below a critical level.

***Exam* tip**: Learn the correct definitions for both long-day and short-day plants, using the term *critical period*.

The principle of negative feedback

Q1 Give the definition of physiological homeostasis.

Q2 Describe what is meant by a receptor in homeostasis.

Q3 Describe what is meant by an effector in homeostasis.

Q4 Describe what is meant by a corrective mechanism in homeostasis.

Q5 Using the terms receptor, effector and corrective mechanism, describe the mechanism of negative feedback control.

ANSWERS

1 The maintenance of an internal steady state within the body of an organism.

2 Receptors contain cells that are sensitive to the levels of particular factors in the internal environment.

3 Effectors are muscles or glands that can bring about corrective mechanisms.

4 A corrective mechanism brings about changes that return a factor to its normal level.

5 In negative feedback control, changes in a factor are detected by a receptor, which triggers an effector to bring about a corrective mechanism. Once the corrective mechanism has returned the factor to normal, the corrective mechanism is switched off.

***Exam* tip:** Learn the definition of negative feedback given in Answer 5. It can be asked as a 2-mark question in your exam.

Osmoregulation

Q1 State the location of the receptor cells responsible for the detection of changes in the water concentration of blood in mammals.

Q2 Name the gland responsible for the production and secretion of ADH.

Q3 Describe the effect of ADH on the kidney tubules.

Q4 Explain the effect of a decrease in ADH on the concentration of salts in urine.

Q5 Describe the effect of a decrease in water concentration in the blood on the release of ADH.

ANSWERS

1 hypothalamus (in the brain)

2 pituitary gland

3 ADH increases the permeability of kidney tubules to water.

4 Decrease in ADH makes the tubule less permeable so less water is reabsorbed. This results in more water in the urine, which decreases the concentration of salts remaining in the urine.

5 Decreasing water concentration triggers the release of more ADH from the pituitary.

***Exam* tip**: Always link the increase or decrease in ADH production with the increase and decrease in the kidney tubules' permeability and the effect this has on the reabsorption of *water*, the volume of urine produced and its concentration.

Glucose regulation

Q1 State the location of the receptor cells that detect blood glucose concentration in a human.

Q2 Name the organ that produces the hormones insulin and glucagon.

Q3 State what effect the detection of a decrease in blood glucose concentration would have on both insulin and glucagon concentrations in the blood.

Q4 Following an increase in its concentration, glucose is removed from the blood and is converted into a storage carbohydrate.

Name the storage carbohydrate and the organ in which it is stored.

Q5 Explain how negative feedback is involved in the homeostatic control of blood glucose concentration.

ANSWERS

1 pancreas

2 pancreas

3 A decrease in blood glucose concentration will cause a decrease in insulin concentration and an increase in the concentration of glucagon.

4 Glycogen, stored in the liver.

5 Changes in glucose concentration from normal are detected by receptors in the pancreas, which produces hormones that act on the liver to correct the concentration. Once the concentration returns to normal the correction mechanism is switched off.

***Exam* tip:** Remember this helpful rhyme.

Low blood sugar, glucose gone. What you need is glucagon.

To turn glucose into glycogen – what you need is insulin.

This is an occasion when spelling is important. You must not confuse *glycogen* and *glucagon*.

Temperature regulation (1)

Q1 State the importance of body temperature to metabolic processes in humans.

Q2 State the exact location of the temperature-monitoring centre in humans.

Q3 State how messages are carried from the temperature-monitoring centre to the effectors.

Q4 Describe the changes that would occur in the skin of mammal as a result of an *increase* in body temperature.

Q5 Describe the changes that would occur in the skin of mammal as a result of a *decrease* in body temperature.

ANSWERS

1 Optimum temperature is required for optimum enzyme activity.

2 hypothalamus

3 by nerves

4 Sweat glands increase activity/increased sweat production, arterioles in the skin vasodilate, hairs lie flat against the skin/hair erector muscles relax.

5 Sweat glands decrease activity/decreased sweat production, arterioles vasoconstrict, hair erector muscles contract/hairs raised.

***Exam* tip:** Candidates often fail to see the link between temperature control and optimum enzyme activity. Make sure you do!

Temperature regulation (2)

Q1 Explain the effect of vasoconstriction on the body temperature of a mammal.

Q2 Explain the effect of vasodilation on the body temperature of a mammal.

Q3 Explain the effect of contraction of the hair erector muscles on the body temperature of a mammal.

Q4 Explain the effect of sweating on the body temperature of a mammal.

Q5 Explain the effect of shivering on the body temperature of a mammal.

ANSWERS

1 Vasoconstriction narrows the arterioles, which decreases blood flow to the skin. This results in less heat being lost from the skin surface by radiation.

2 Vasodilation widens the arterioles and increases blood flow to the skin. This results in more heat being lost from the skin surface by radiation.

3 Contraction raises the skin hairs, which trap a layer of air close to the skin surface. The air forms an insulating layer that decreases the heat loss from the skin surface.

4 Sweat moistens the skin surface. The liquid sweat uses heat from the skin surface as it changes into a gas (evaporates), so cooling the skin.

5 Muscle contraction of shivering generates heat.

***Exam* tip:** Candidates often lose marks here because of lack of detail, description and the correct adjectives and terminology. Learn the responses given here.

Endotherms and ectotherms

Q1 Give the term used to describe an organism that regulates body temperature through changes in its behaviour.

Q2 Describe what is meant by the term endotherm.

Q3 State the effect of an increase in environmental temperature on the metabolic rates of both ectotherms and endotherms.

Q4 Explain why an endotherm's metabolic rate is greater when its surroundings are at 10 °C than when they are at 30 °C.

Q5 Describe a behaviour in an ectotherm that is likely to raise its body temperature.

ANSWERS

1 ectotherm

2 An animal that obtains most of its heat from its own metabolism and whose body temperature is held constant in spite of changes in the external environmental temperature.

3 Increases the metabolic rate in ectotherms. The metabolic rate of endotherms is unaffected.

4 Metabolism rises to compensate for greater heat loss to surroundings at 10 °C.

5 Basking in direct sun *or* sitting on hot objects such as rocks.

***Exam* tip:** You must be able to state how endotherms and ectotherms gain their heat and maintain their body temperature.

Factors affecting population change

Q1 Explain what is meant by density-dependent and density-independent factors that affect population change.

Q2 Name *two* density-dependent factors that may affect the numbers in a wild population.

Q3 Name *two* density-independent factors that may affect the numbers in a wild population.

Q4 Describe how a density-dependent factor affects population density.

Q5 Describe how a density-independent factor affects population density.

ANSWERS

1 The effect of density-dependent factors is intensified as the population density increases. The effect of density-independent factors is unaffected by the density of the population.

2 Food supply, disease, predation, and competition for space, minerals, water or light in plants (any two).

3 Extremes of temperature, extremes of rainfall, e.g. floods or drought.

4 As a density-dependent factor increases, its effect on the population increases, which in turn reduces the effect of the factor on the population.

5 Density-independent factors act on populations in a way that is unaffected by the density of the population.

***Exam* tip:** Pay extra attention to Answers 1, 4 and 5 – once again it is the language required that allows candidates to gain the marks.

Monitoring populations

Q1 Populations of animals are relatively stable despite short-term oscillations in number.

Give the meaning of this statement.

Q2 Wild populations are monitored to collect essential data because of the organisms' importance as sources of food or raw materials.

State *three* other reasons why wild populations are monitored.

Q3 Explain the use of indicator species.

Q4 Explain what is meant by an endangered species.

Q5 Describe why a species might be described as a pest.

ANSWERS

91 Monitoring populations

1 The population numbers may fluctuate/change/increase or decrease due to food availability, disease, predation and weather, but over a longer period the numbers remain relatively stable.

2 To produce data on pollution, endangered species or pest populations.

3 The presence or absence of a certain species can indicate the conditions in the environment *or* the abundance of an indictor species is related to the concentration of a polluting substance in a habitat.

4 An endangered species is at risk of extinction.

5 A pest species has a detrimental effect on humans or their activities.

***Exam* tip**: Remember that it is the purpose to which monitoring data is put that is important.

Succession in plant communities

Q1 Give the term used to describe the changing sequence of plant communities in a habitat over time.

Q2 Succession is unidirectional.

Explain what is meant by this statement.

Q3 Suggest a modification made by one plant community to its habitat that favours colonisation by another community.

Q4 State the effect of succession in plant communities on biomass and the complexity of food webs.

Q5 Name the final community in a normal succession sequence and describe the features of that community in terms of species diversity, biomass and the complexity of food webs found within it.

ANSWERS

92 Succession in plant communities

1 plant succession

2 It always occurs in the same direction through time.

3 Improved soil fertility, humus, soil depth, pH, drainage or water retention.

4 Biomass and food web complexity are increased during succession.

5 Climax community, has a high diversity of species, complex food webs and high biomass compared with previous communities.

***Exam* tip**: It is the changes that a community makes to the habitat which are important for marks. Make sure you can give the characteristics of the climax community.